Baby's
OUTER SPACE

Black and White
High-Contrast Book

Written and illustrated by

R. M. Smith

CLARENCE-HENRY BOOKS

Blast,
zoom,
rockety
rocket ship!

Hello,
floaty
astronaut!

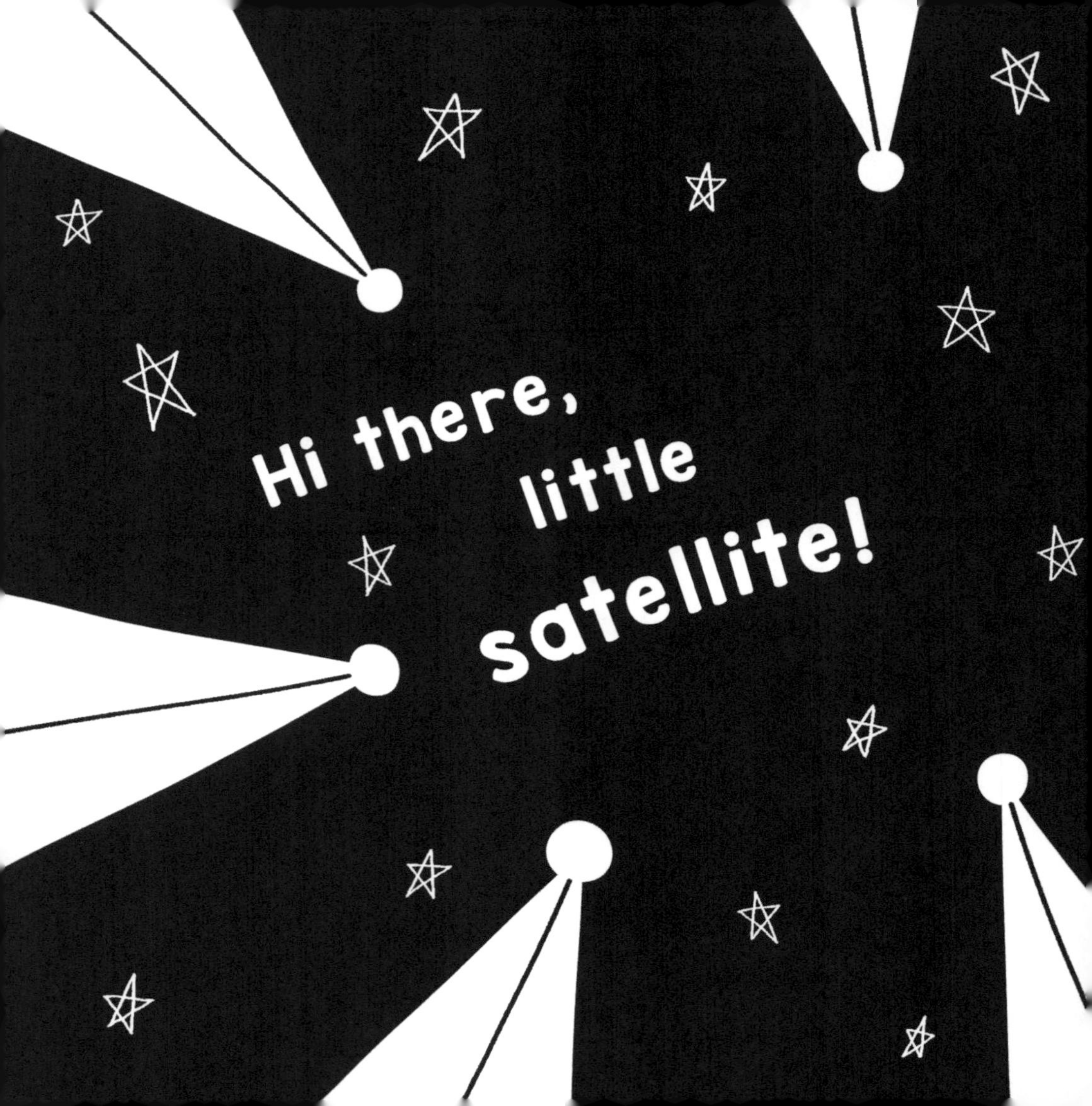

Hi there,
little
satellite!

Glowing bright,
round **moon!**

Good to see you,
friendly telescope!

Looking bouncy,
lots of planets!

Hello twinkling, shining stars!

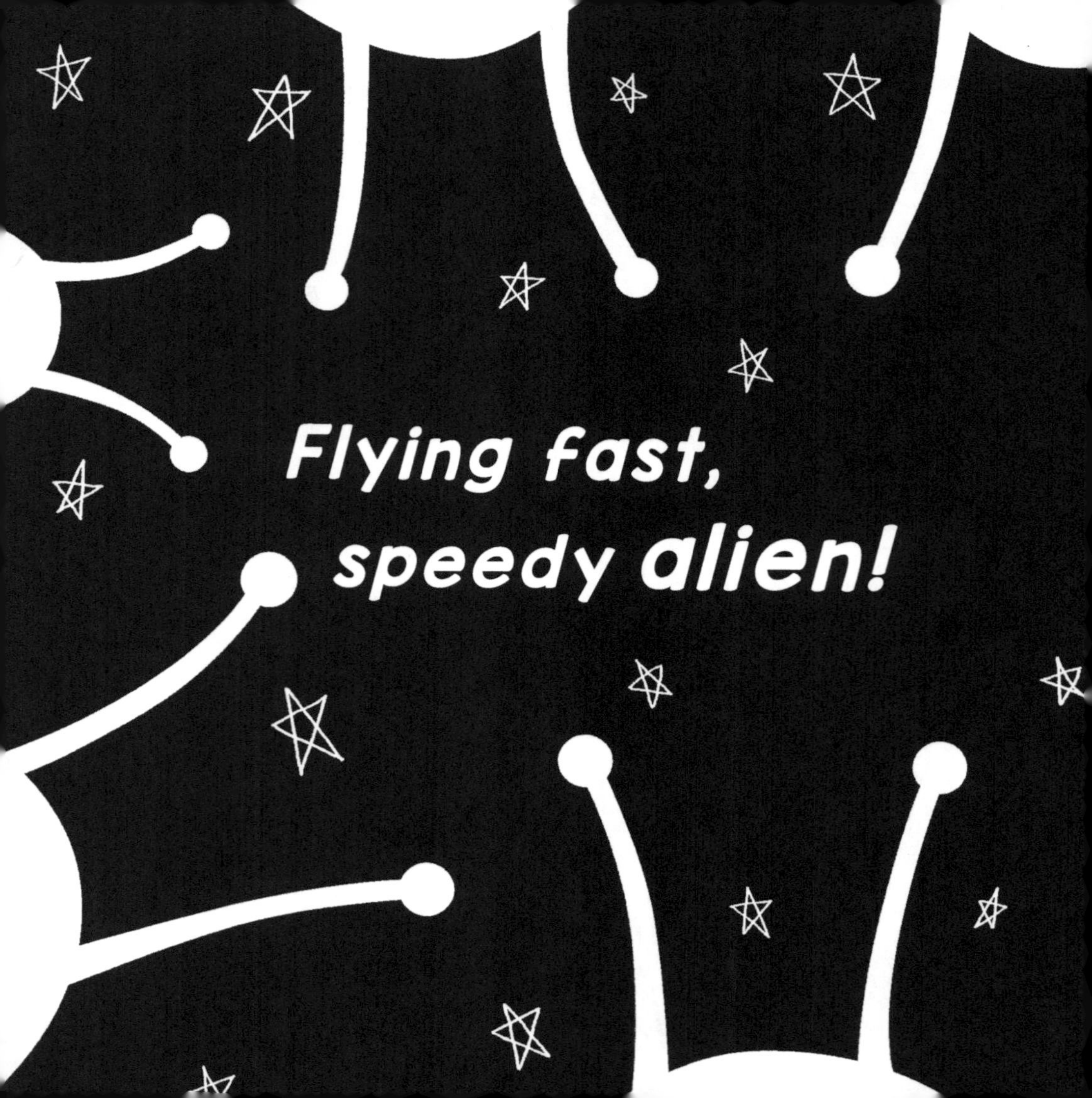

**Flying fast,
speedy alien!**

Hey there, big bear!

Whirl around, giant galaxy!

Whoosh,
whoosh,
icy comet!

What's up,

big bright

sun?

Keep on smiling, happy Earth!

Baby's
OUTER SPACE
by R. M. Smith

Clarence-Henry Books • Alexandria, VA
Copyright © 2024 R. M. Smith

Design and Layout by R. M. Smith

Summary: A high-contrast black and white baby book
featuring objects from outer space.

ISBN-10: 0988290960
ISBN-13: 978-0988290969

First Edition
10 9 8 7 6 5 4 3 2 1